Flat Earth? Round Earth?

Flat Earth? Round Earth?

Written and Illustrated by

Theresa Martin

Prometheus Books
59 John Glenn Drive
Amherst, New York 14228-2197

Published 2002 by Prometheus Books

Flat Earth? Round Earth? Copyright © 2002 by Theresa Martin. All rights reserved. No part of this publication may be reproduced, stored in a retrieval system, or transmitted in any form or by any means, digital, electronic, mechanical, photocopying, recording, or otherwise, or conveyed via the Internet or a Web site without prior written permission of the publisher, except in the case of brief quotations embodied in critical articles and reviews.

Inquiries should be addressed to
Prometheus Books
59 John Glenn Drive
Amherst, New York 14228–2197
VOICE: 716–691–0133, ext. 207
FAX: 716–564–2711
WWW.PROMETHEUSBOOKS.COM

06 05 04 03 02   5 4 3 2 1

Library of Congress Cataloging-in-Publication Data

Martin, Theresa, 1963–
Flat earth? Round earth? / written and illustrated by Theresa Martin.
p. cm.
ISBN 1–57392–988–3 (alk. paper)
1. Earth—Figure. I. Title.

QB631 .M33 2002
525'.1—dc21

2002021991

Printed in the United States of America on acid-free paper

For Martin,

who loves seeking the answers
as much as asking the questions

Flat Earth? Round Earth?

In science class, Mrs. Markum gave us balls of light-weight clay to make models of the earth. We had blue clay for the oceans and gray clay for the continents. When Stan took his ball and smashed it flat, we all laughed.

"What's so funny?" Mrs. Markum asked, but then she saw his flattened earth and her lips got tight. "Why did you destroy your model?"

"I didn't," Stan replied. "I made it flat, the way the earth really is."

Mrs. Markum opened her mouth and shut it, and the rest of us stood waiting.

"That's silly," she finally said. "You know that the earth isn't flat."

"It is so flat," Stan said. "Look outside the window."

We all turned our eyes to the window, even though we already knew what we would see. Our playground with its swings and slides ended in a barbed wire fence. Beyond that the prairie rolled away to the horizon.

"It just looks flat because it's so big," Mrs. Markum said. "We can't see the curve because we're so very tiny."

"But if it was round, the people on the bottom would fall off."

"You know why they don't fall off," Mrs. Markum snapped. "It's because of gravity."

Stan shook his head.

"What do you mean, no?" Mrs. Markum demanded. "You've seen pictures of Earth from space. It's a sphere, like an orange. Everyone knows that."

"It is not," Stan said. "And you can't make me believe it." He turned his back on her and walked away.

So Mrs. Markum sent Stan to the principal's office, and the rest of us went back to work molding the clay into continents.

But I was bothered. I knew that the earth was round. Everyone did. That's what I learned in science class. We had a globe at home with all the continents on it. I'd seen pictures taken from outer space, and diagrams of the solar system.

But Mrs. Markum said that the scientists who change the way we think about the world are the ones who ask questions and don't just take everyone else's word for how things are. Maybe Stan was right and the rest of us were wrong.

I shouldn't simply accept Stan's idea of what the world looks like, any more than Stan should accept Mrs. Markum's. In science class we learned that ideas like Stan's and Mrs. Markum's are called "hypotheses." A hypothesis is a guess about what the world is like. Because Stan and Mrs. Markum have different hypotheses, we need to look closely at the earth itself to see which hypothesis is more likely to be true. If the earth is flat, like Stan says, then we should find more things about the world that show it to be flat, but if the earth is round, then we'll discover more facts that make us think it's round. So I ought to be able to prove it, one way or the other. But how?

I walked over to Stan's empty desk and brought his smashed earth over to my desk. I set it down next to my round earth. Dad told me that a long time ago everyone thought the world was flat and if you sailed a ship far enough, you'd fall right over the edge. I imagined the frightened seamen sailing farther and farther away from shore, sure that at any moment their ship would drop over the edge of the world.

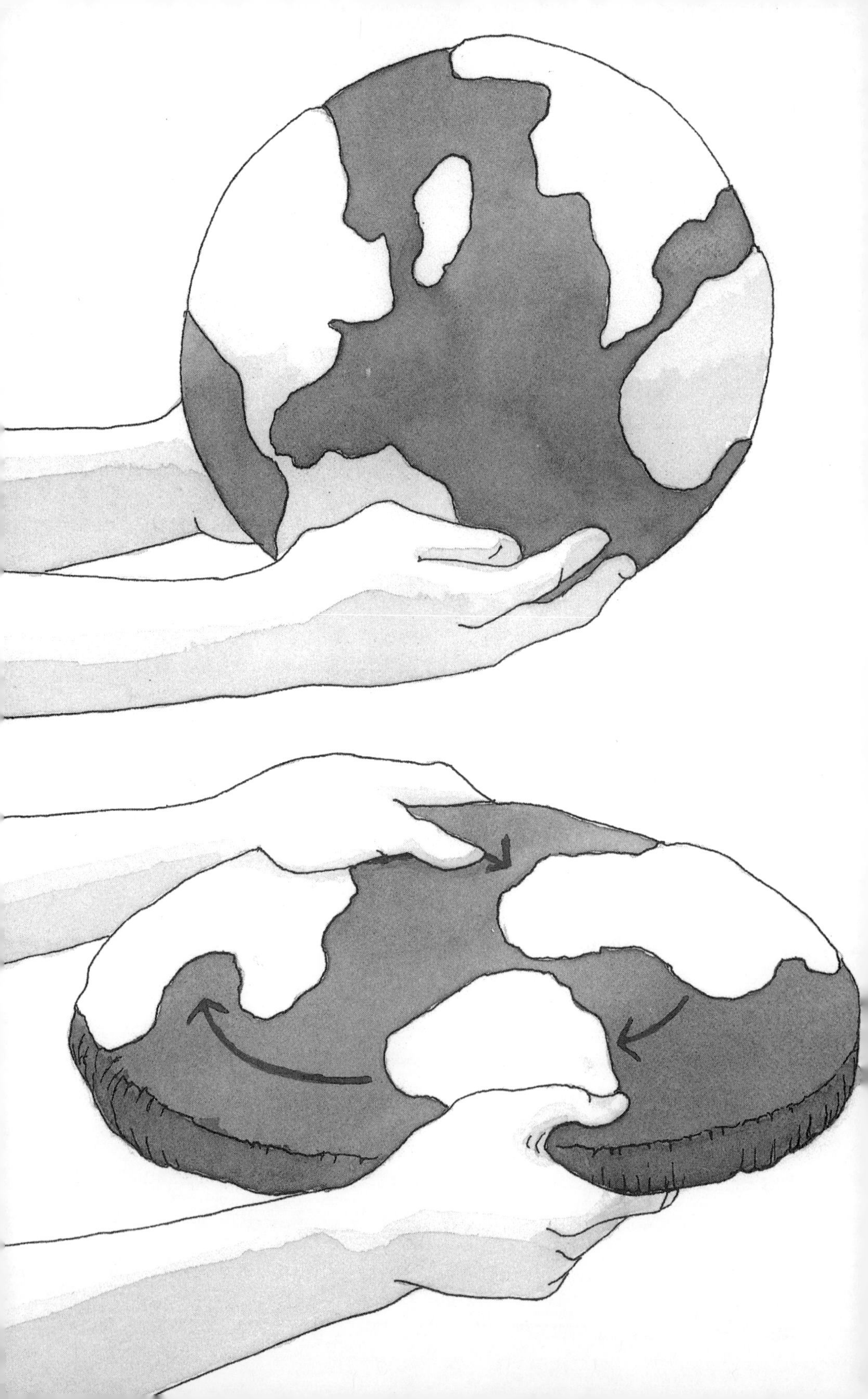

That never happened. I lifted up the round earth. There was Europe where Christopher Columbus started. He sailed across the Atlantic Ocean and discovered the Americas. Now ships left Europe, traveled across the Atlantic Ocean, went through the Panama Canal or around the tip of South America, and crossed the Pacific Ocean only to hit Asia. And Asia was connected to Europe, so it was possible to travel by ship all the way around the globe. I traced the whole route with my fingers, gouging a little path through Central America for the Panama Canal, which I had forgotten to put in.

Then I picked up Stan's flat earth and tried to do the same thing. It didn't work. Even if all the continents were on the same side of the disk, you'd have to travel in a circle rather than a straight line to hit all those continents.

I wanted to show Stan. Surely he could see that he was wrong and not make any more trouble for himself. Since it was time for recess, I shoved both balls under my shirt. When Mrs. Markum led the class down the hall, I slipped around the corner to the office.

Stan was slumped on a bench outside the principal's office. His science notebook was clutched in his lap.

"Hey, Nathan," he said.

"Did you get in trouble?" I asked.

He shook his head. "I haven't seen him yet. What are you doing here?"

"I want to show you something," I said. "Maybe you won't get in trouble."

"What?"

I took out both balls and sat down next to him.

"Here are two models, a flat earth and a round earth. If I built a ship, and set sail here from the East Coast of North America and always kept my compass aimed east, see how I could go all the way around?"

"It's not always the same direction," Stan pointed out.

"No, I'd have to do some turning to get through the Mediterranean and Red Seas, but I could adjust for that, and for going through Indonesia. But see, I'd end up back here on the West Coast of North America."

"How do you know?" Stan asked. "Have you ever done it?"

"No," I said, "but lots of people have."

"How do you know? Maybe they just made up those stories."

"But I *could*."

"But you haven't."

"Okay, okay," I said. "I'll think of something else."

I sat for a long time staring down at the two balls, one round and the other flat. Inside the office I could hear the secretary's fingernails clicking on her keyboard, but my eyes never left the smooth blue curve of the ocean on my ball. The overhead light reflected off the top, and then it curved away from me until it disappeared down toward my knees. That was it!

"Have you ever been to the ocean?" I asked.

"No," Stan said. "Have you?"

"Yes," I cried. "I went to Cape Cod last summer. We watched the ships from Boston Harbor go through the Cape Cod Canal on their way out to sea."

Stan was staring at the balls. I didn't know if he was listening.

line of sight
ship sinking below horizon

"Do you know what happens as they get farther and farther away?"

"No," he said.

"They get smaller, of course, because everything looks smaller when it gets farther away. But do you know what else?"

"No."

"They look like they're sinking down. They don't just get smaller and smaller until they're so tiny they disappear. They look like they're dropping down below the horizon. Here, I'll show you." I handed Stan his flattened earth and bent down to look underneath the bench. The only thing there besides dust was a piece of an eraser. I picked it up and set it upright on the edge of North America on my round earth. Then I squeezed a little piece of modeling clay off the edge of Stan's flat earth and set it flat next to the upright eraser bit.

"The eraser is me and the bit of clay is the ship." I held the whole thing up in front of his face. "See how when the ship travels away from me, it goes around the earth and looks like it's sinking lower?"

Stan wrinkled his brow and frowned, so I asked him if I could borrow a piece of his notebook paper, and I drew him a diagram.

"See? Doesn't that make sense?"

"Yes," he said slowly. "But maybe it's just my eyes playing tricks on me."

"What do you mean?" I asked.

"Like that magician when he came to do a magic show last year. He made it look like his head was floating and it wasn't. He called it an 'optical illusion.' That's when you think you see something, but it isn't real. He said that's what it is when the sun looks extra big when it first comes up over the horizon in the morning."

That gave me an idea.

"Forget the ships," I said. "Here's another one. What are you doing when the sun comes up?"

"Depends on the time of year," he said.

"Well, how about now?" It was late May, so the sun rose pretty early.

"Doing chores, usually," he said.

"Think about how it gets light. You can probably see a long way from your farm."

"All the way to the next town," he replied proudly. "We can see the tops of the grain elevator from the hayloft door."

"And it's pretty flat?"

"There are a few dips and hollows, but overall, it's pretty flat. Good grazing country."

"Think about when the sun first comes up," I said, leaning forward. "Does the sunlight hit everywhere all at once, or does one thing get light before the rest?"

His brow wrinkled again and his lips pursed, but he wasn't frowning this time.

"Even when the yard's still dark," he said slowly, "the sun hits the top of the grain elevators. Then it hits our barn."

"How does everything else get light?" I asked.

The furrows in Stan's brow deepened. "You know, when I'm up in the hayloft, I can see light sweeping from the east across the prairie toward the grain elevators. Once it starts, it's really fast. It takes less than a minute to go from our yard to the grain elevators."

"Bingo!" I cried.

"What?"

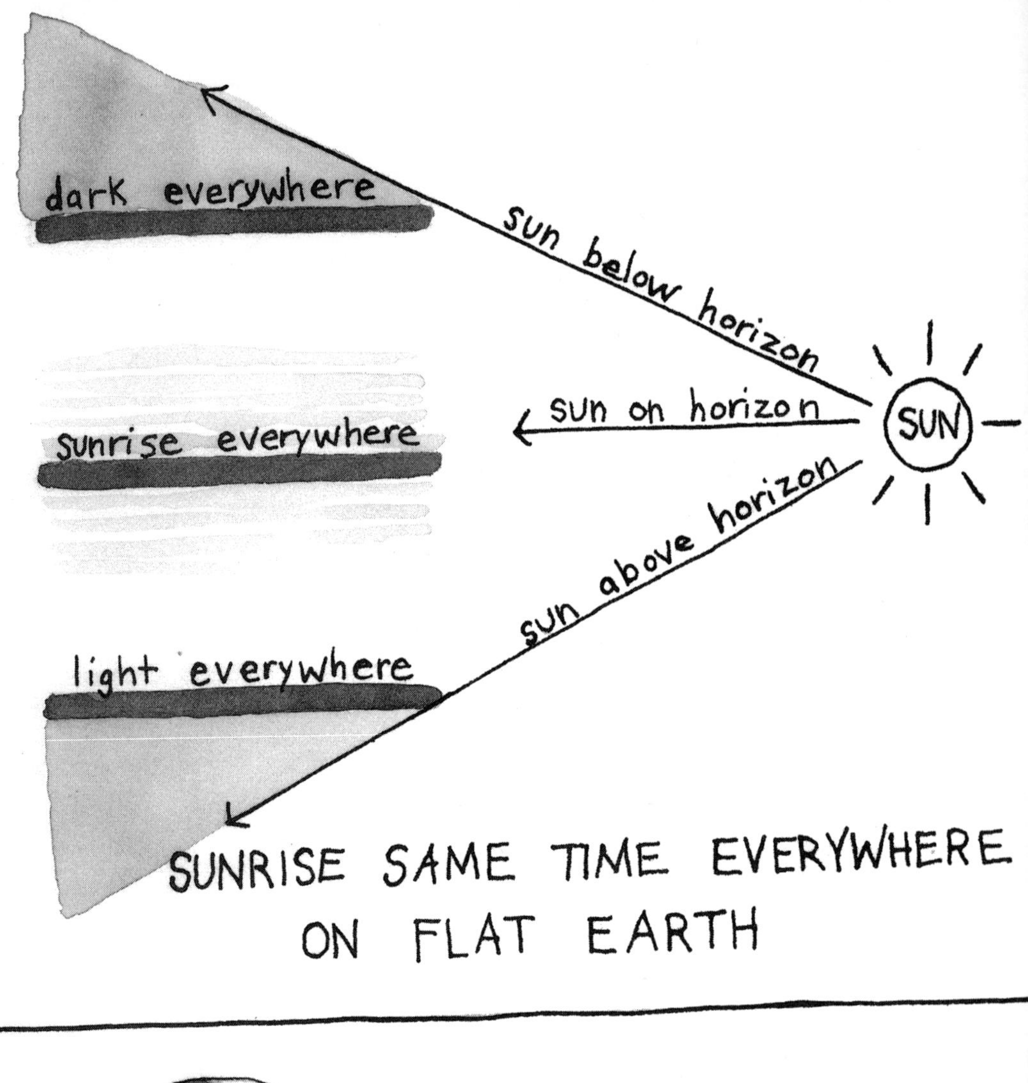
dark everywhere
sun below horizon
sunrise everywhere
sun on horizon
SUN
light everywhere
sun above horizon
SUNRISE SAME TIME EVERYWHERE
ON FLAT EARTH

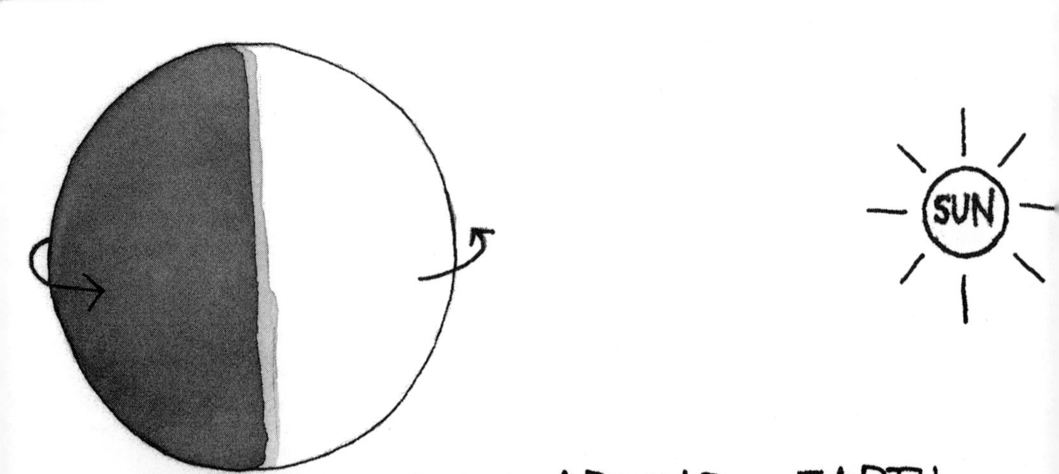
SUN
SUNRISE MOVES AROUND EARTH
AS ROUND EARTH SPINS

"The sunlight moves across the prairie because the earth is round. You saw it yourself. The sun comes up around the earth and the line of light moves as the sun comes up higher reaching more of the rounded globe."

"I don't get it," Stan said. He was frowning now.

"Here, look," I said. I lifted up his flat earth. "If the earth were flat, then when the sun rose, the light would hit everywhere at once." I then picked up the round earth. "And if the earth were round, then the light would have to move around it. The sunlight hits the top of the grain elevator first because it's so high up. But since you say the light moves across the plains to get there, it has to be round."

Stan's frown deepened. "No it doesn't," he said. "There are a lot of little bumps on the earth, like for mountains and stuff. The plains could just be part of a curve here."

"They're not just rounded here," I said, but then I realized that I wasn't really thinking.

San Francisco
Denver
Kansas City
Washington D.C.

"When the sun comes up, the line of the sunrise moves across the earth. We've both seen it. But how can we know that it's not something that happens just here?" I was thinking out loud. "If the earth is round, the sunrise would be a bit later the farther west you go. Hey, I bet if we got a paper from Denver, the time of sunrise would be just a little bit later than here, and if we got a paper from San Francisco, it'd be even later. And if we got papers from Washington, D.C., and Kansas City, they'd be earlier. There's no way that would be true if the earth were flat."

"How are you going to find that out?" Stan asked.

"At the library," I said, leaping to my feet. "I bet I can get there and back before recess is over."

I peeked into the office and saw only the secretary's back, so I slipped out the front door and ran down the block to the public library. A rack held newspapers from across the country. I pulled down the *Denver Post*, *San Francisco Chronicle*, *Kansas City Star*, and *Washington Post*. Those cities were all basically along the same line going east and west. It took a while to find where they listed sunrise times, but I found them. I was tempted to tear out that part of each page so that Stan couldn't say I made up the times, but I didn't because someone else might need to look them up, too.

San Francisco — Denver — Kansas City — Washington DC

Washington DC	5:46 Eastern Time	− 1 hour	4:46 Central Time
Kansas City	5:56 Central Time	→	5:56 Central Time
Denver	5:35 Mountain Time	+ 1 hour	6:35 Central Time
San Francisco	5:50 Pacific Time	+ 2 hours	7:50 Central Time

Stan was still sitting on the bench when I got back. He was staring up at the light on the hall ceiling, and only looked down when I shoved the piece of paper with the sunrise times in front of his face.

"See, it works!" I declared.

He squinted at the numbers, wrinkling his brow and frowning. "But the times farther east aren't earlier at all," he said.

"You have to account for the different time zones," I said. "Denver is in Mountain Time, so you add one hour, and San Francisco is in Pacific Time so you add two hours. And here's Washington, D.C. You subtract one hour from that because it's in the Eastern Time zone. The sun rises first in Washington, D.C., then in Kansas City, then Denver, and last in San Francisco. Pretty neat, huh? We could even graph it! What do you think, now?"

Stan scratched his head.

"Don't you see that the earth has to be round, if it takes all that time for the sun to rise from one side of North America to the other?"

San Francisco
Denver
Kansas City
Washington DC
T. Martin

"I don't know," he said. "It could be that North America is one big curve. It does rise up from sea level to pretty high in the middle. That could be why it takes so much time for the sun to rise the farther you go." He drew a picture of it on a sheet of notebook paper. I tried not to let my jaw stiffen into what I knew was a stubborn look.

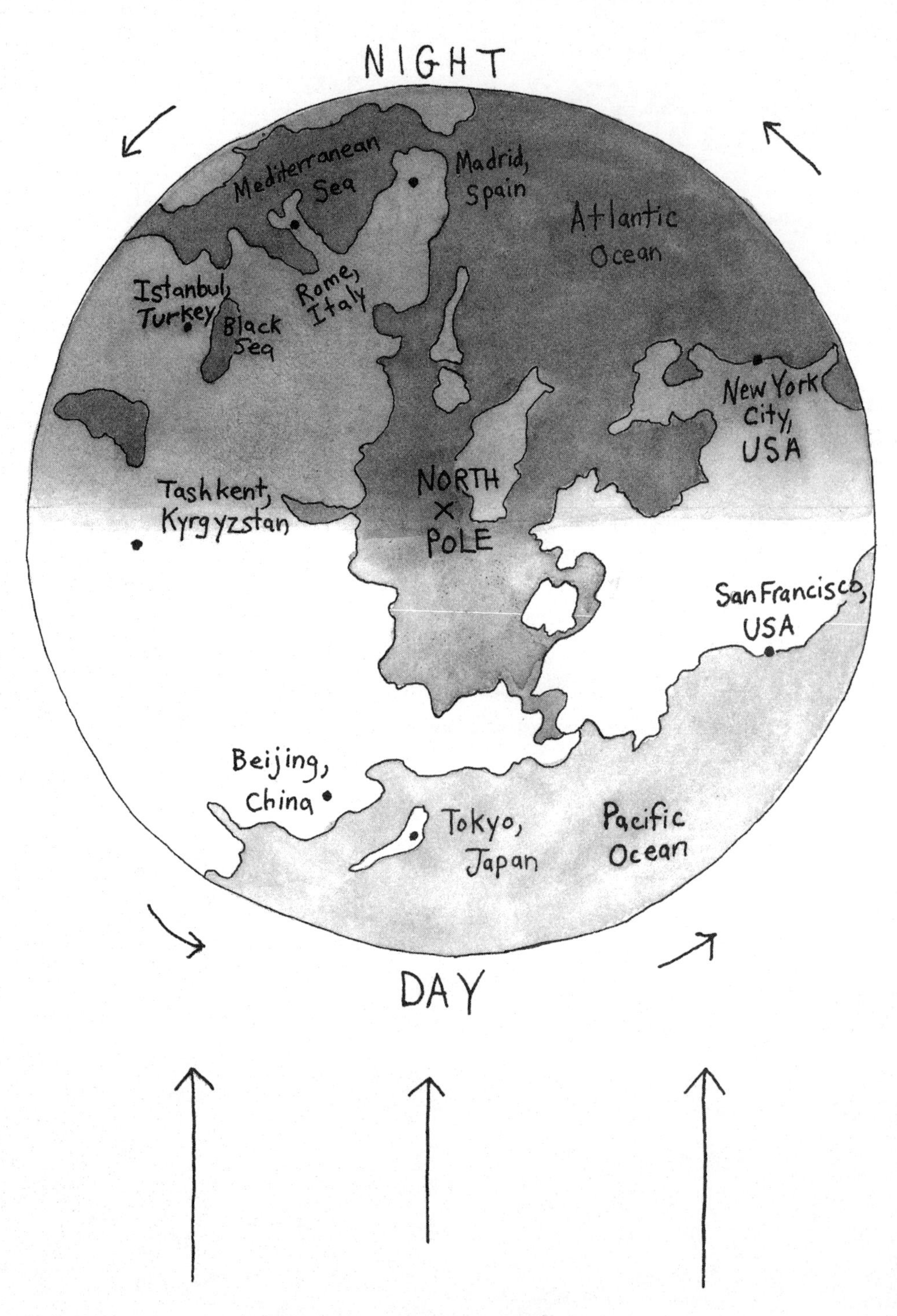
NIGHT
Mediterranean Sea
Madrid, Spain
Atlantic Ocean
Istanbul, Turkey
Rome, Italy
Black Sea
New York City, USA
Tashkent, Kyrgyzstan
NORTH POLE
San Francisco, USA
Beijing, China
Tokyo, Japan
Pacific Ocean
DAY
SUN SHINING

"If you admit that it's rounded in North America, why shouldn't it also be for the rest of the world?"

"It's just not."

"But I bet we could trace the sunrise times in Europe and Asia and the Pacific Islands."

"How could you do that? We don't have those newspapers here in town."

"I guess not," I said. "But we could if we lived in a city that had them, or we could find them on the Internet."

"You still haven't proved anything," Stan said quietly, letting his hands rest on his knees and his eyes drift back up to the light on the ceiling. There was some kind of dead bug stuck to it. "You should go back to class. I heard them come in from recess a little while ago."

"No, I'm going to show you that the earth is round." I glanced up and down the hallway. I didn't have any arguments left.

← sunlight
90°
90°
90°
90°

I looked down at the two models Stan had put next to him. I saw the shadow on the bench that my round earth made.

"Okay, Stan, I'm going to draw you a picture." I drew a wide curve on the bottom half of a sheet of paper. Then I drew a sun far up at the top. I drew three little spikes pointing straight out of the curve of the earth.

"If you could call up three friends—one who lives a thousand miles straight north, one who lives five hundred miles north, and one who lives five hundred miles south—and tell them to take a yardstick outside and stick it straight up and down—and have them mark on the ground where the end of the shadow would be, those shadows wouldn't be the same length if the earth were round."

"Why not?"

"Because the sun wouldn't be at the same angle everywhere. The farther north you go, the less overhead the sun is."

Stan frowned at the picture.

"But you'd get different shadows with a flat earth, too."

"How do you figure?"

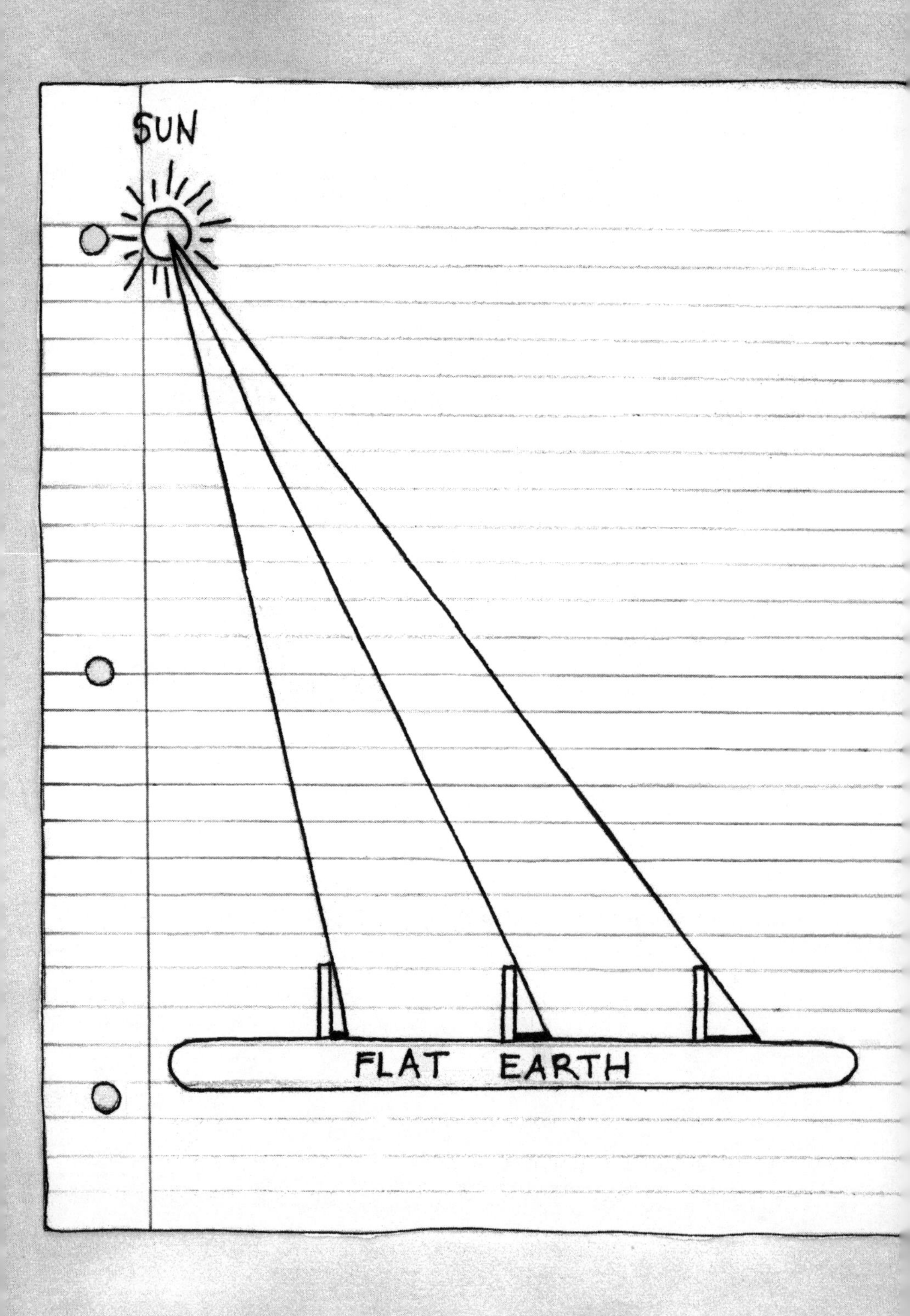
SUN
FLAT EARTH

Stan drew a straight line on the bottom of another piece of paper and a sun at the top. He drew three short vertical lines across the bottom and drew lines from the sun to cross the tops of each vertical line. "See, these make different angles, too."

"The sun is really much, much farther away than that," I pointed out. "It's so far away that these angles would all be the same."

"Maybe it's really close," he said, drawing the sun even closer. It made each angle even more different. He smiled at me. I didn't smile back. I stood up and picked up my modeling clay earth.

"I think you want to get in trouble," I said.

"No, I don't," Stan replied. "I just don't think you're right."

How could I not be right? Here was the earth. I lifted it up over my head and made it travel a slow curve through the air in the hallway. It was beautiful up there with its shiny blue oceans and bumpy gray continents. A movement on the floor caught my eye. It was the shadow of the ball crossing the round school logo painted on the floor inside the doorway.

"An eclipse!" I shouted.

"What?" Stan looked up. The typing in the office stopped.

"You've seen an eclipse, haven't you?" I asked.

"Sure," Stan replied.

The secretary appeared in the doorway. "What's going on out here?" she asked.

"I'm showing Stan how we know the earth is round."

"Of course it's round. You boys know that," she said, and disappeared back into the office. I didn't even look at her.

"During an eclipse, when the shadow of the earth blocks out part of the sunlight reflecting off the moon, what shape is the shadow?"

"It's curved," Stan said.

"And for a total eclipse?"

"I don't know. I never stayed up that late."

"It's round," I said. "I've seen it."

Stan shook his head, but I slapped my hand against my thigh. "That proves it! Watch the shadow on the floor." By raising it above my head, I could make the shadow of my earth big enough to fill the round logo on the floor. "It has to be round."

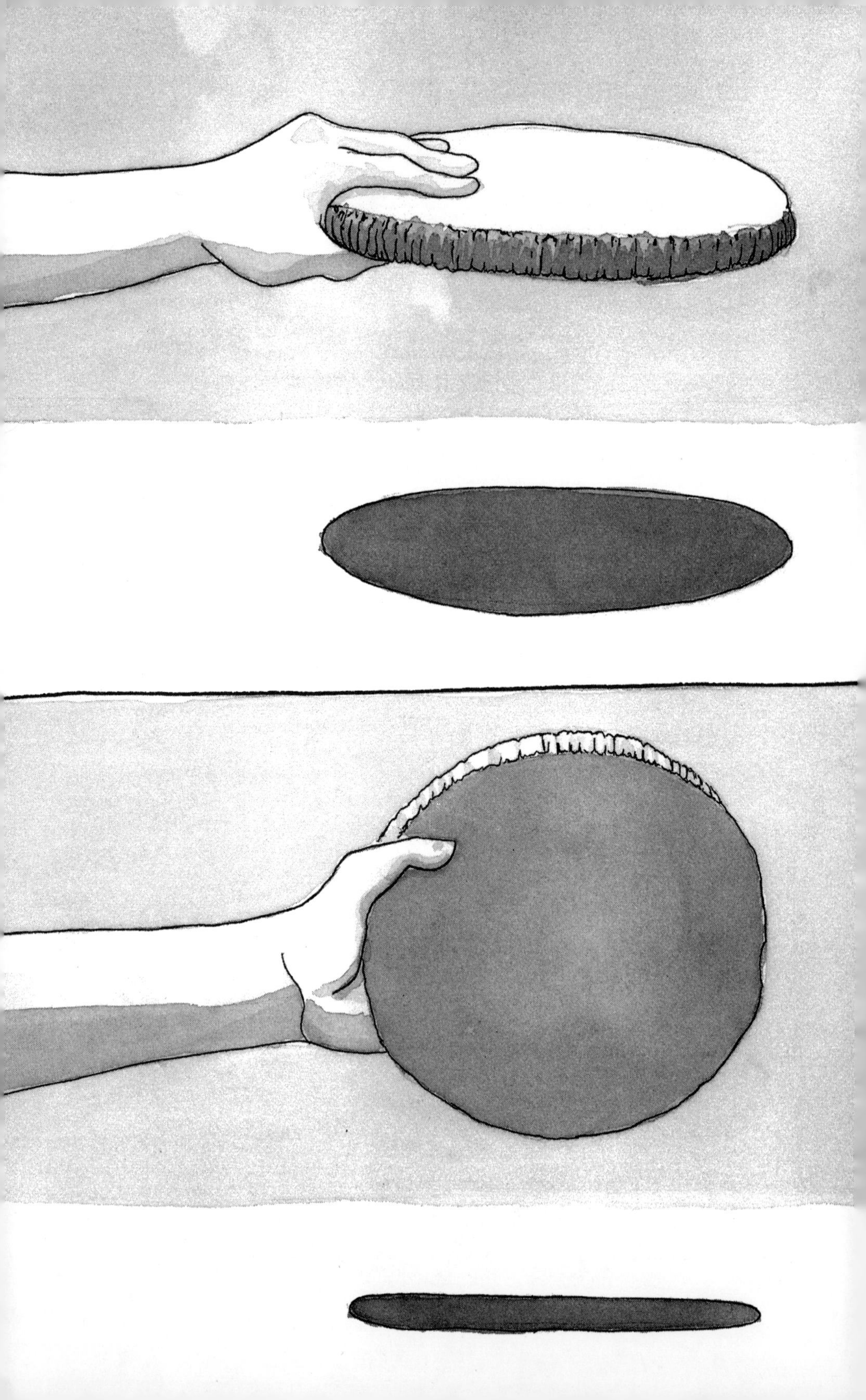

"But I can do that, too," Stan said, standing up and passing his flat earth under the glare of the overhead light. It also made a round shadow.

"But what happens if it isn't lined up perfectly?" I said. I grabbed it away from him and turned it so that it was sideways to the light. The shadow was now sausage-shaped. "Only with a sphere is the shadow always a perfect circle. Eclipses never make sausage-shaped shadows."

"So maybe the flat earth is always lined up perfectly."

"Why should it line up perfectly?" I asked.

"Why shouldn't it?"

I had no answer. I looked at Stan's smug face and suddenly hoped that the principal would yell at him.

	OBSERVATIONS SUPPORTING A ROUND EARTH	STAN'S EXPLANATIONS
1.	Traveling around globe	Made-up stories
2.	Ships sinking over horizon	Optical illusion
3.	Moving lines of light/darkness	Bumps in earth's surface
4.	Later sunrise times across continent	Continent is one huge bump
5.	Different shadow lengths in north and south	Different angles of nearby sun
6.	Curved shadow on eclipse	Perfect lining up of disk-shaped earth

"I'm going back to class now," I said. "Good luck."

"Thanks," Stan said. "I know you tried. I just don't happen to believe you. You haven't told me anything that I can't find another answer for."

"But all your answers are different," I protested. "Everything I've shown you supports the hypothesis that the earth is round. Your explanations all say something different."

"So what?" Stan said.

"If I had to choose one explanation or the other," I said, "I'd choose the one that explains the most."

"Why?"

"Because that's how you test a hypothesis. The more different observations you make to support a hypothesis, the more likely the hypothesis is true. Nothing we've talked about disproves it, and everything supports it."

The secretary came out of the office again. "Mr. Carlson will see you now, Stan," she said. Stan got up and followed her without even looking back.

I walked back to my classroom. Everyone was bent over composition books.

"Where have you been, Nathan?" Mrs. Markum asked me.

"Trying to convince Stan that the earth is round," I said.

"I'll have to give you detention for not coming back to class after recess," she said.

"Okay." I could always work on my model some more. I sat down at my desk and got out my composition book. The assignment was written on the board. "Imagine that you are Christopher Columbus. What are you thinking as you set sail for India?"

I started writing, but then felt a hand on my shoulder.
It was Mrs. Markum.
"Did you convince him?" she asked.
"No," I said. "But I convinced myself."